BEARS

BEARS

TOM JACKSON

This pocket edition first published in 2024

First published in a hardback edition in 2020

Copyright © 2024 Amber Books Ltd

All rights reserved. No part of this publication may be reproduced, stored in a retrieval system, or transmitted in any form or by any means, electronic, mechanical, photocopying, recording, or otherwise, without prior written permission of the copyright holder.

Published by
Amber Books Ltd
United House
London N7 9DP
United Kingdom

www.amberbooks.co.uk
Instagram: amberbooksltd
Pinterest: amberbooksltd
Twitter: @amberbooks

ISBN: 978-1-83886-358-6

Project Editor: Anna Brownbridge
Designer: Keren Harragan and Rick Fawcett
Picture Research: Terry Forshaw

Printed in China

Contents

Introduction — 6

Types of Bear — 8

Habitats — 70

Family — 138

Cubs — 188

Picture Credits — 224

Introduction

Bears blend great strength and ferocity with gentleness and contentment. At least, that is our perception of these mammals, which are the largest terrestrial carnivores on the planet. If we were to meet one in the wild, our focus should be on the bear's ability to injure or kill us. Its huge, rounded body is not for cuddling but is a platform for a powerful musculature that can be almost effortlessly violent. Of course, the world's bears, from the cute panda to proud polar bear, are not bloodthirsty man-

eaters. They would prefer to be left alone by us – and by each other. The sun bears of Asian rainforests are the archetypal honey lovers, while the sloth bears of India like nothing more than to slurp up ants. Contrary to expectation, black bears are not always black and polar bears are not white. What other surprises await? Let's begin an exploration of the life of the bear.

ABOVE:
Watch and observe
As with all bears, a polar bear cub learns its valuable life lessons by watching mother.

OPPOSITE:
Need for protection
Most bear species, especially the pandas, are threatened by human activities.

Types of Bear

There are eight species of bear in the world today. They form the family of animals called the Ursidae. Together they evolved from dog-like ancestors about 30 million years ago. Despite the difference in scale, a similarity in the shape of face and skull is clear. The most widespread and numerous species is the brown bear, which is found across the northern hemisphere, mostly in the colder regions but historically ranging as far south as North Africa. It is well named because the English word 'bear' is itself derived from the old Germanic word for 'brown', and the brown bear's binomial name, Ursus arctos, comes from the Latin term for bear, ursus, plus the Greek term arctos, which also means 'bear'.

The brown bear is in the subgroup, or genus, of species that includes the polar bear, Ursus maritimus (meaning 'sea bear'), the American black bear, Ursus americanus, and Asian black bear, Ursus thibetanus. Members of the Ursus genus tend to be bigger than the remaining four species, all of which occupy their own branches of the bear family tree. The sloth bear, Melursus ursinus, and sun bear, Helarctos malayanus, of southern Asia are the closer cousins, while South America's spectacled bear, Tremarctos ornatus, is somewhat isolated from the rest. The most oddball family member is the giant panda, Ailuropoda melanoleuca, which was once conjectured not to be a bear at all, but a giant raccoon.

LEFT:
Sun bear
Among its many names, this species is also called the dog bear for its small size (about 1m, or 3.3ft, long) and short, glossy coat.

RIGHT:
Sun sign
On lifting the chin to show its throat and chest, the sun bear reveals the source of its name. Each bear has a sun-like ring (or portion thereof) of flaming golden fur.

OPPOSITE TOP:
Diet
Sun bears are mostly fruit eaters, exploiting their small size to clamber through trees to reach them. When fruit is unripe and hard to find, the bear will supplement its diet with insects gleaned from branches, plus roots and leaves.

OPPOSITE BOTTOM:
Lone ranger
Similar to all bears, sun bears want to be alone, left to their own devices to forage in peace in their own patch of territory. When two do meet, there is invariably a show of strength and then a swift departure from the weaker individual.

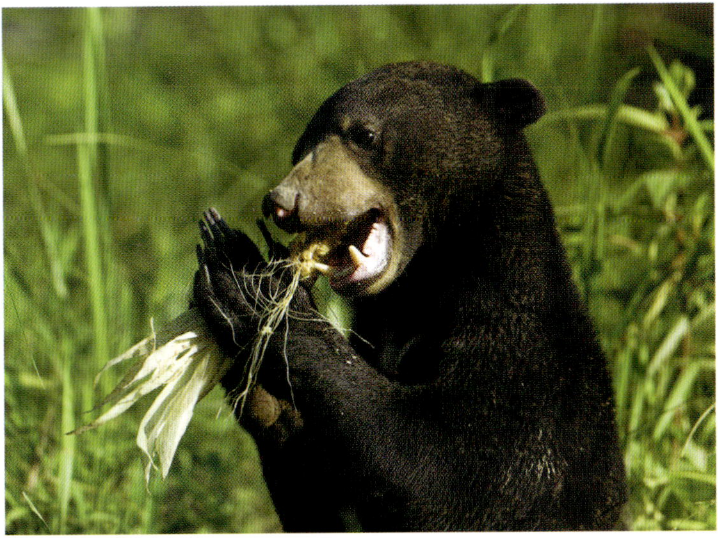

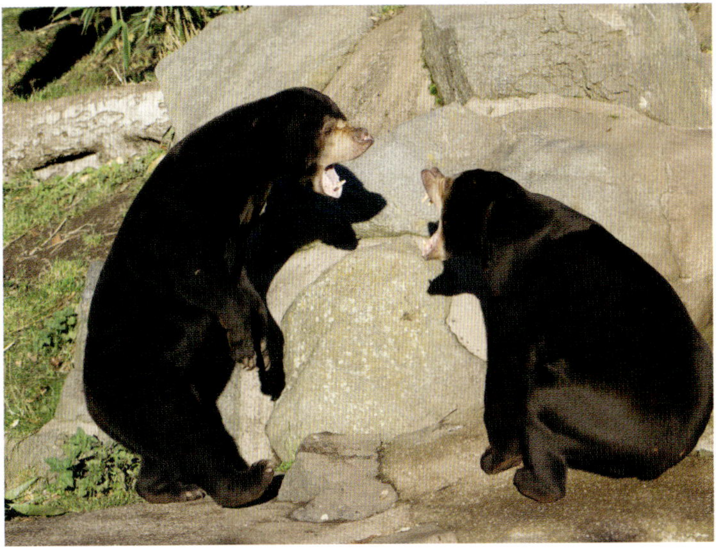

OPPOSITE:
Paw licker
A sun bear likes to feast on ants and termites and uses its claws to dig a little way into the nest. Out swarm the defending insects on to the bear's mighty paws, whereupon the larger animal slurps up the little ones with its long tongue.

LEFT:
Full height
Bears are famous for standing upright on their back legs, although they do not walk on two feet in the wild (that is a trick taught to performing bears). Standing helps bears to assess each other's size and identity – and intimidate foes.

ABOVE:
American black bear
American black bears have perhaps the most catholic tastes of all bears. This cub has been given a salmon to feast on but will also eat roots and berries.

RIGHT:
Tree climbers
With 95 per cent of its diet destined to be plant based, this American black bear cub is doing well by perfecting the skill of climbing into trees to find food.

RIGHT TOP AND OPPOSITE:
Warmer location
Although the American black bear shares some of its range – at a distance – with the larger brown bear, it also occupies warmer areas to the south. Unable to sweat like humans do, it sheds excess heat by panting.

RIGHT BOTTOM:
Cool off
In the often oppressive heat and humidity of the Deep South of the United States, American black bears try to reduce their body temperature by taking a dip in shallow swamp water and natural pools.

Brown bear
The brown bear is easily differentiated from its diminutive cousins by the bulge above its shoulders. This is from the immense shoulder blades and neck vertebrae required to anchor the muscles needed to hold up the bear's heavy head.

RIGHT:
Cold climate
In the main, the brown bear is at home in a cold and wet habitat, such as the forests of Canada and Siberia.

OPPOSITE:
True monster
The brown bear competes with the polar bear for the crown of largest land mammal and mightiest terrestrial carnivore. Males are a fifth bigger than the females, and can be 3m (9.8ft) from nose to tail – and taller still when stretching upright to show off their full size.

OVERLEAF:
Small eyes
Although the giant head makes its facial features appear small, a bear's eyes are still not large and do not offer very acute vision. The long-sighted bear will rely more on smells and sounds instead.

Big mouth
A large male brown bear can grow large enough to weigh up to 680kg (1500lb), although most weigh much less. The largest bears of all are on the Alaskan island of Kodiak, where the salmon runs in spring are an important source of food for the bears as they emerge from their winter sleep.

OPPOSITE:
Thick fur
Brown bears have a thick double-layered coat. Against the skin, there is a short underfur that insulates the animal. Over the top there are longer, coarser guard hairs, coated with oils to create a waterproof outer layer, which stops dirt and bugs getting to the underfur.

LEFT TOP:
Omnivore
Despite their reputation as a killer, bears are omnivores, meaning 'all eater'. They hunt prey, scavenge carcasses, glean insects, search out fruits, fungus and roots, and will even eat leaves and grasses to add weight before winter.

LEFT BOTTOM:
Eurasian subspecies
Brown bears in Europe and Asia are not as large as their American cousins. They were once common across Europe and Asia but are now mostly restricted to Siberia and Scandinavia.

Claws

All bears have long, curved claws. They are non-retractable, meaning they are always exposed. Despite looking like a weapon, the claws are more often deployed in climbing and digging.

RIGHT:
Grizzled
The Canadian subspecies of brown bear, *Ursus arctos horribilis*, is also known as a grizzly. This is due to its grizzled hair, where paler tips on the brown shaft give a silvery colour to the bear's coat.

OVERLEAF:
Making a spectacle
Named after the way the pale facial markings often form a ring around each eye, the 1.5m (4.9ft) long spectacled bear is the second-largest animal in South America after the tapir.

Forest bear
Also known as the Andean bear, the spectacled bear is the most arboreal, or tree-based, member of the bear family, spending a large part of its time up trees, where it searches for its favourite meal: fleshy epiphytes.

OPPOSITE:
Unique markings
The pale rings or arcs around the eyes of this species are unique to each spectacled bear and are used to identify particular individuals, much like humans do with our fingerprints.

LEFT TOP:
A day's sleep
The bear appears to be active at all times of the day and night – and during twilight hours – and so will catch some sleep in a secluded spot, such as a comfy tree nest, whenever it fancies.

LEFT BOTTOM:
Long legs
The spectacled bear is the last surviving relative of the extinct short-faced bears. These giant creatures had long legs for galloping over prairie land to chase down prey. The Andean bear survives in a very different way but retains long legs and a flat face.

Sea mammal
By rights, the polar bear should not be classified as a terrestrial hunter at all and should be included as a marine mammal instead. The bear spends most of its time out at sea, in the water if necessary, but mostly patrolling the ice pack in search of prey.

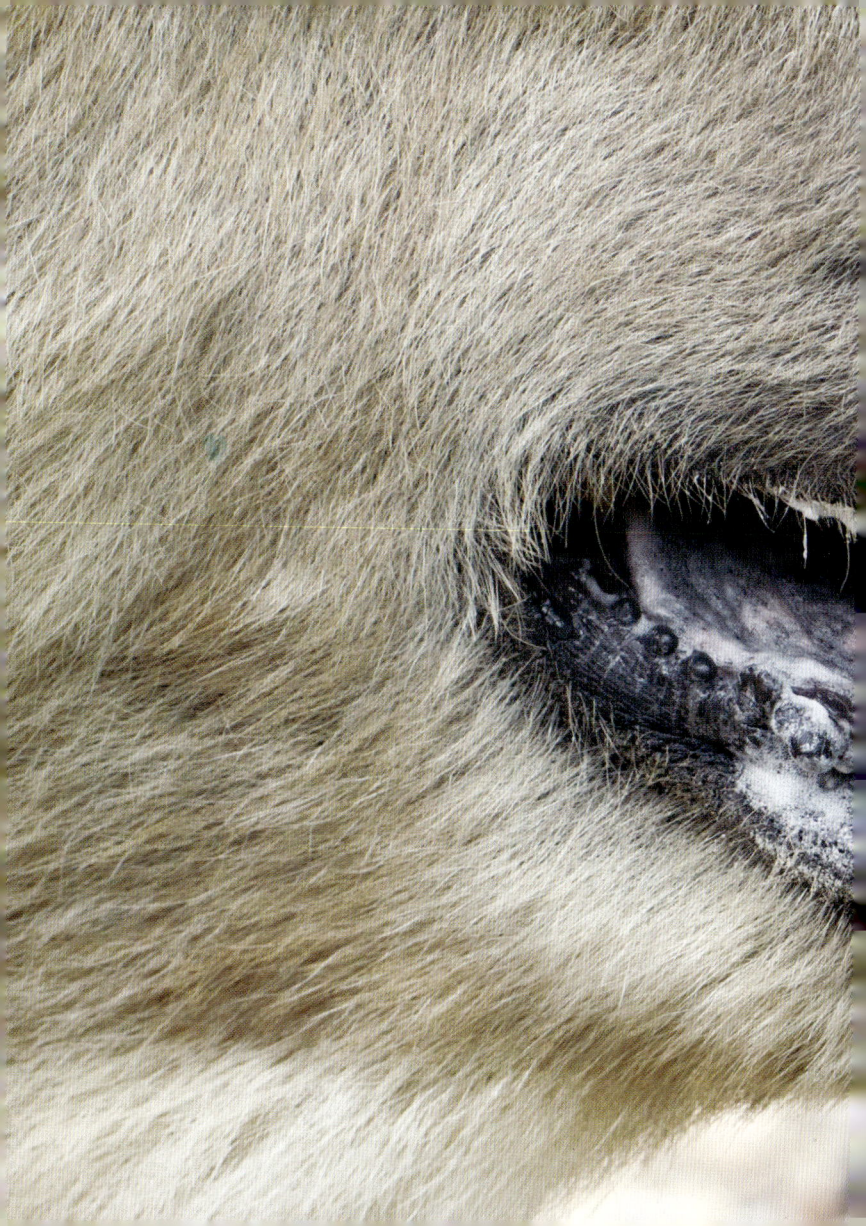

Slicing teeth
The dentition, or arrangement of teeth, of bears reveals their dietary needs. Instead of the scissor-like cheek teeth of dogs and cats, the bears have more rounded teeth better suited to grinding up plants than slicing flesh. The polar bear relies on the immense strength of its bite to chew up meaty meals.

RIGHT:
White colour
What colour is a polar bear? The obvious answer is white but that would be wrong. In fact, the polar bear's hairs are transparent, with a hollow centre where the pigments would ordinarily go. Together, the colourless hairs reflect light, which is why they appear white.

OPPOSITE TOP:
Flat feet
Bears have a plantigrade stance, which means the toes and heel bones are in contact with the ground at the same time. This provides a stable platform for walking over slippery or loose ground. One of the few other animals to share this unusual stance is humans.

OPPOSITE BOTTOM:
Insulation
The shafts of the polar bear's transparent hairs are filled with air, which is a poor conductor of heat. Therefore, the hollow hairs add to the insulation provided by the bear's fur and fat.

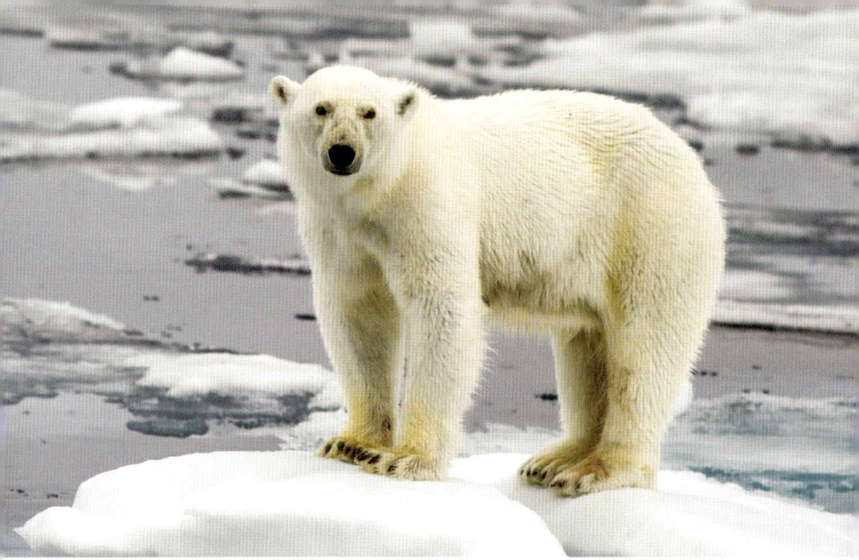
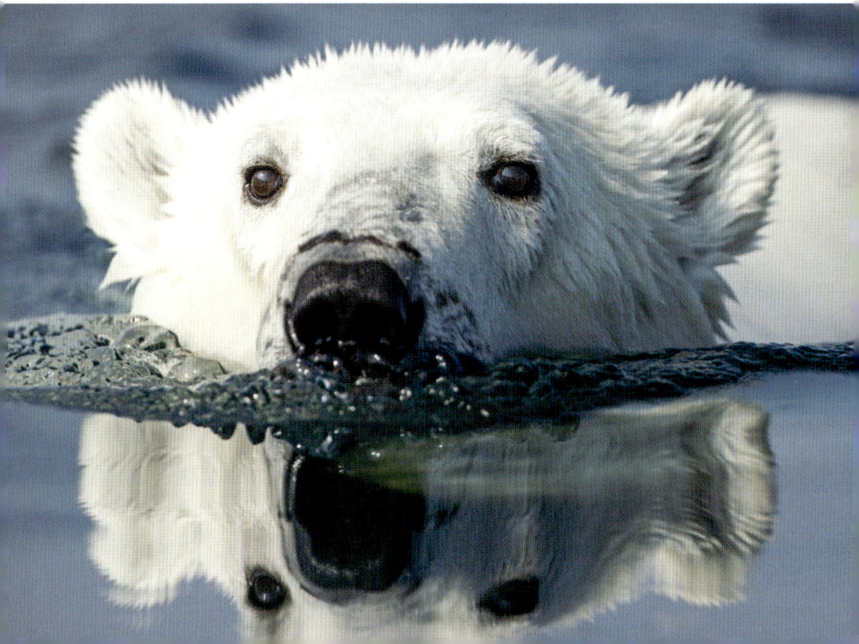

Covering ground
Polar bears spend the spring and summer on the move, in constant search of food. Prey is spread thinly in the polar desert of the Arctic and some bears have to patrol vast home ranges of 300,000 square kilometres (116,000 square miles) to find enough to eat.

Winter sleep

The polar winter arrives quickly and lasts for several months. For weeks on end there will be no sunlight at all and the temperatures will drop to below -40°C (-40°F). Even a polar bear cannot survive this, and enters a dormant state – although it is not technically deep enough to be a hibernation. Males simply fall asleep under the snow, whereas pregnant females will dig a larger den into the ice.

RIGHT TOP:
Air warmer
Polar bears have a slightly arched snout compared with those of other species. Inside are plates of thin bones that create a labyrinth inside the large nasal cavity, which warms the cold in-breaths before they hit the bear's lungs. (All bears have this but not to the extent of the Arctic species.)

RIGHT BOTTOM:
Diving
A diving polar bear can hold its breath for two minutes. The bears are active swimmers and can be found far out from the edge of the ice in summer. They are often on the move to a better hunting ground, and are not equipped to catch prey at sea.

OPPOSITE:
Paddle power
The polar bear has wider feet than the average bear. They act like snow shoes in that they spread weight on the snow so the bear does not sink too deeply.

RIGHT:
Asian black bear
A close cousin of the American black bear, the Asian black bear lives a similar life, but in the forests of East Asia and the Himalayan foothills. One distinction from the American species is that Asian black bears have a denser ruff of hairs on the neck, which makes the bear look bigger and meaner than it really is. This is possibly an adaptation to the threat of tiger attacks, something the American black bear does not face.

ABOVE:
Moon mark
The Asian black bear is also called the moon bear, a happy coincidence with its continental neighbour the sun bear. The name comes from the pale crescent markings on the chest, which the bear displays by standing up.

RIGHT:
Under threat
Asian black bears are vulnerable to extinction because they have been hunted for their paws, claws and gallbladders, which are used in folk medicine. The bears are now protected apart from in Japan, where they can be shot legally in wild areas.

ABOVE:
Mountain men
The Asian black bears that live on the Tibetan plateau are probably the chief inspiration for the Yeti myth, also known as the Abominable Snowman. Bears might also have been the source of such American monster myths as Bigfoot and the Sasquatch.

RIGHT:
Slurp
Asian black bears use their acute sense of smell to locate beetle grubs up to 1m (3.3ft) under the ground. They dig down with their powerful forelegs and slurp up the nutritious treat with their long and agile tongue.

RIGHT:
Face myth
Few animals are as recognizable as the giant panda. Legend has it that the bear was once white all over but when mourning the death of a young shepherd girl killed by a leopard, the bears adopted the human custom of covering their arms with dark ashes. Then wiping away their tears and covering their ears to block out the wails of despair led to the distinctive facial markings.

RIGHT TOP:
Playtime
A panda cub learns climbing skills. Like most of the smaller bears – the largest pandas grow to about 80cm (2ft 7in) to the shoulder – this species spends part of its time in trees. Pandas do this to avoid predators, although few natural threats remain.

RIGHT BOTTOM:
Nap time
In summer, the Chinese bamboo forests can become sweltering and the pandas cool off by resting on a tree branch. They only sleep four hours at a time, because they must keep eating to fuel the large body.

OPPOSITE:
Living alone
The black and white features of the panda are the opposite of camouflage. Instead of trying to stay hidden in the forest, the bear is easy to see through the greenery, and so the chance of an unwanted close encounter is reduced.

OVERLEAF:
Captive breeding
For many decades, the giant panda has been classified as endangered, with numbers perilously low. A worldwide captive breeding programme in zoos and wildlife parks (mostly in China, where the panda originates) has managed to boost numbers, and now the species' survival prospects have been upgraded.

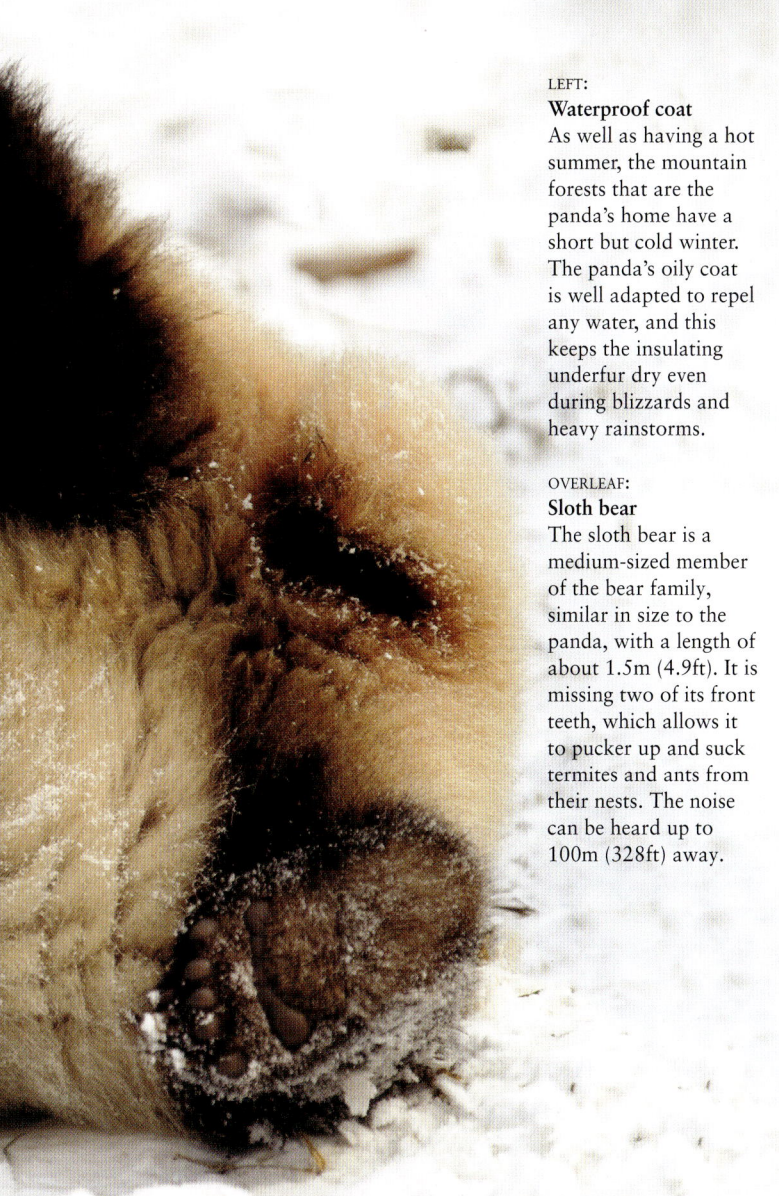

LEFT:
Waterproof coat
As well as having a hot summer, the mountain forests that are the panda's home have a short but cold winter. The panda's oily coat is well adapted to repel any water, and this keeps the insulating underfur dry even during blizzards and heavy rainstorms.

OVERLEAF:
Sloth bear
The sloth bear is a medium-sized member of the bear family, similar in size to the panda, with a length of about 1.5m (4.9ft). It is missing two of its front teeth, which allows it to pucker up and suck termites and ants from their nests. The noise can be heard up to 100m (328ft) away.

RIGHT TOP:
Ant-eater
The sloth bear is mostly found in the Indian subcontinent. It survives on a restricted diet of ants, termites and fruits. Unusually for bears, sloth bears may gather in groups around a large food source, only tolerating each other while the food supply remains plentiful.

RIGHT BOTTOM:
Shaggy hair
The sloth bear has very long and shaggy hair compared with other bears. This helps with making the bear seem larger when meeting predators or rivals. The coat may also protect the skin against pests and protect against the strong sunlight in open habitats.

OPPOSITE:
Flexible lips
The sloth bear's lips are highly flexible so they can pluck the choicest berries from the thorniest bush. An earlier name for the species was the 'lip bear'.

RIGHT:
Entertainer
The sloth bear is the most common species used as a performing bear by Indian street entertainers. Although this practice is now illegal and much less common, the dancing bears were invariably captured as cubs and had claws and teeth removed.

OPPOSITE:
Back claws
The common name of 'sloth bear' comes from the way this species has long claws on its hind paws almost matching those on the forepaws. This allows the bear to hang upside down from branches, for a short while at least, in the pose more associated with the sloths of the Americas.

Habitats

Bears are mostly northern animals, attuned to life in the cold and wet. The eight species live on all continents bar Australia and Antarctica, but only two – the sun bear and spectacled bear – are found south of the equator.

Today, all species of bear are under threat, attacked on all sides by the activities of their greatest enemy: humans. For centuries, bears have been persecuted, shot on sight by farmers who worry they are a danger to livestock and by villagers fearing they threaten human lives. As a result, bears have lost their footing in all but their stronghold, the cold boreal forests of North America and Asia, where few people seek to compete with them. This is where brown and black bears live in their greatest density. The north truly is the land of the bear. It is no coincidence that the word Arctic is derived from the Greek term arctos, meaning 'bear'.

Beyond cold, wet conifer forests, bears have spread further north on to the frozen tundra and sea ice; this move was perhaps as little as 70,000 years ago, meaning polar bears are very close cousins to the brown bear. Black bears have moved south to warmer deciduous woodlands. Pandas are famously limited to the bamboo forests of China. The sun bear lives in steamy tropical jungles in Southeast Asia, while the sloth bear is found in drier monsoon forests and shrublands in the Indian subcontinent. Spectacled bears live furthest south of all but only in the cool heights of the Andes Mountains.

LEFT:
Winter arrives
Bears are built to put on weight as a thick layer of fat under the skin. When the cold winter comes, this fat supply will keep the animal alive as it sleeps through the extreme weather.

Seal hunt
A female polar bear has successfully tracked and killed a ringed seal. This prey will be a much-needed meal for her and her twin cubs, who need to grow large enough in time for the winter fast approaching.

RIGHT TOP:
Honey
A black bear braves the stings of bees to smash up their nest and get at their honey and larvae. The bees cannot sting through the bear's fur but will attack the exposed skin on the lips, nostrils and around the eyes.

RIGHT BOTTOM:
Taste for bamboo
Pandas only ever eat bamboo, a thick, fast-growing relative of grasses. The bear crushes the stalk and chews up the fibres. The only other food is the occasional insect or bug that is inside the bamboo.

OPPOSITE:
Leaving a mark
To advertise their presence, bears scrape tree trunks with their claws to leave a marking. This is also often anointed with the bear's scent by rubbing its oily fur on it. Tree-scraping also helps to keep claws sharp and healthy.

Home ranges

Bears do not defend strictly demarcated territories. Instead, their home ranges will overlap with neighbouring ones. This female brown bear will lead her cubs around an area of 100 to 1000 square kilometres (39 to 386 square miles), while the local males will roam over twice that area and overlap her territory.

PREVIOUS PAGE:
Sea ice
The polar bear's habitat is sea ice, a pack of ice that is up to 3m (9.8ft) thick in some places. By late summer, however, the ice pack is thinning out into fragments of drift ice.

RIGHT:
Autumn urgency
This polar bear appears very at home in the Arctic waters bathed in low autumn sunlight and surrounded by the slush of a rapidly freezing ocean. As the sea ice returns in October, the bear will have only a few weeks left to fatten up for winter.

OPPOSITE:
Fish supper
The river systems that feed into the North Pacific are full of wild salmon and, each spring, as the adult fish swim upstream to spawn, the brown bears of Canada and Siberia, still ravenous after their winter starvation, gorge themselves on fish.

LEFT TOP:
Grab and swipe
The bears catch fish by batting them from the water with their hefty forepaws. A single blow is enough to kill or stun the fish. Fishing like this takes some skill and bears are most successful in the dark, using touch and hearing to guide them to the tasty target.

LEFT BOTTOM:
Bonanza
Brown bears can catch a fish every seven minutes, about eight every hour. With such a success rate, they bite out the most nutritious parts, such as the liver and heart, before discarding the hard-to-eat head and tail.

Sixth finger

Giant pandas have evolved a sixth finger to help them grip the stalks of bamboo. More accurately described as a pseudothumb, it is a spur of bone that extends from the wrist and acts like an inflexible version of an opposable thumb.

Night and day
The sun seldom sets in the short Arctic summer, and so by default the polar bear is a diurnal creature, forced to stalk its prey in the clear light of day. All other bears are mostly nocturnal, although they will feed during the day in periods of plenty.

Too hot to handle
Built for the freezing Arctic, the polar bear can get overheated quite easily. To cool down, it will stretch out on the colder ground to shed its excess heat through its belly.

RIGHT TOP:
Sea foods
This brown bear's home range takes in the coast of Alaska, so it enjoys a meal of seaweed and barnacles on the rocky shoreline.

RIGHT MIDDLE:
Clam dig
This grizzly knows that when the tide goes out it can find clams buried in the mud. Cracking open the shells will not be a problem for this mighty beast, although digging them up might be quite a messy business.

RIGHT BOTTOM:
Digging in
It might look like this sloth bear is taking a bite of wood. However, it is using its teeth and claws to dig into the dead wood to get at the beetle grubs that are burrowing around inside.

OPPOSITE:
Branch-bender
The spectacled bear has a debonair look as it perches in the bow of a tree. The bear uses its long legs to reach out to flimsy branches and pull them and the fruit they hold within reach of its mouth.

Dustbin raider
The wilderness areas of North America have a strong attraction for tourists wanting to experience the great outdoors. The food these people bring to the camp, and the waste they leave behind, is also a great attraction for wild bears.

Paddler
Although it is much more agile walking on four legs, a polar bear can swim for several hours at a speed of around 10km/h (6.2mph).

LEFT:

Underwater search
Brown bears find food in water by snorkelling. The bear swims at the surface with its eyes and snout in the water. It is searching for dead fish that have sunk to deeper parts of the river bed, which will make a good meal that is easier to catch!

BELOW:

Buoyancy aids
With their layer of subcutaneous fats and thick fur set up to hold countless pockets of air, bears are born with a natural life jacket. As a result, they find swimming easy, but diving down underwater does not come so naturally.

ABOVE:
Meadowland
As the most widespread species, brown bears have adapted to the widest range of habitats. They were once found in the dry Atlas Mountains of North Africa but are probably extinct there now. Today, beyond the conifer forests, brown bears mostly live in alpine meadows and coastal grasslands.

RIGHT:
Tree refuge
Being smaller and more agile than brown bears, black bears are able to climb higher into trees. This skill is thought to be a hangover from when black and brown bears shared more of their range, and the smaller bears needed to make an escape from their competitors.

RIGHT TOP:
Stalk
Polar bears have two basic hunting techniques. With the stalk, they track the movements of prey (mostly seals on a beach), edging closer along the coast and then stopping to avoid being seen. The final strike over that last 30m (98ft) or so is a thundering charge at up to 55km/h (34mph).

RIGHT BOTTOM:
Still
On the sea ice, seals spend most of their time in the water, but they need to come to the surface to breath at holes in the ice pack. The bears seek out a breathing hole, where they sit or stand and wait, perfectly still. The bear can smell the seal as it approaches under the ice, and will lunge forward and snatch up its victim as it pokes out its head. The bears can even smash through the ice and grab the seals that lurk beneath unawares.

Making a den

Bears generally sleep during the day, when they curl up at the base of a tree in the middle of their home range. As winter approaches, they will prepare a more secluded spot suitable for a stay of several months.

RIGHT TOP:
In the shallows
This Alaskan grizzly bear has driven a sockeye salmon into the shallows of the mountain river, making for a much easier kill.

RIGHT BOTTOM:
Ambush point
Waterfalls are the ideal location for a spot of salmon fishing as the fish gather at the foot of the falls to build up their strength ready for the great leap upstream. When they are ready to go, the bears will be waiting.

OPPOSITE:
High altitudes
The bamboo forests pandas call home need a lot of rain. During the last ice age, much of China that was left ice-free was covered in bamboo, but over time this was replaced by broadleaf trees. Only the tall mountains of southern China have the right cold, wet and often snowy climate for this unique but dwindling habitat today.

In torpor
It is often said that bears hibernate for the winter. In fact, the bears that live in tropical areas, such as the sun bear and sloth bear, are active all year around. Other bears do sleep through the winter, perhaps more than half of the year in more extreme climates, and they survive on the fat stores that were built up in summer. However, the sleep is not deep enough with the attendant drop in metabolism to be a true hibernation.

Race against the weather
A mother and her two cubs, both about two years old, take a last hunting trip as the weather turns ugly in the Arctic autumn. The trio will den over the winter together, and then the younger pair will be ready to start life alone in the spring.

ABOVE:
Spring arrives
Cubs are born in the middle of the winter sleep, when the mother is still fast asleep. This schedule gives cubs time to put on weight on their milk diet before the spring arrives. The heat of the mother's body, which stays high in winter (unlike the body temperatures of true hibernators), keeps babies warm as they grow thicker fur ready for life outside.

OPPOSITE:
Arctic menu
Polar bears have to work hard and cover ground to find enough food, but the prey on offer in the snow and ice is all high-calorie fare, and much needed to survive the long months of winter. The main prey is ringed seals but the bear also kills walruses, beluga whales and sea birds, all of which use fats under the skin as insulation against the cold water.

PREVIOUS PAGE:
Alpine habitat

Spectacled bears explore part of their habitat, here an alpine meadow above the treeline in the Andes Mountains.

RIGHT:
Fruit-picker

Bears have a gap between the canine teeth and their molars called the diastema. This comes in handy when picking berries. The fruit-laden stalk slots into their diastema, and the bear uses the teeth either side to strip off the sweet treats.

FAR LEFT:

Protected areas
Six of the eight species of bear are endangered by habitat loss and human activities. Only the brown bear and American black bear are not globally threatened, and this is mostly due to large areas of wilderness in the United States and Canada being classified as protected areas, where the bears are just as at home as the human residents.

LEFT:

Forest refuge
The spectacled bear once lived from above the treeline down to sea level in the deserts along the Pacific coast, but human encroachment has since meant the bears are now mostly confined to mountain forests, where they clamber into branches to find fleshy plants to eat.

Short summer
In the Arctic summer, there is a rapid reduction in sea ice, so polar bears head inland in search of food. They may end up 100km (62 miles) from the shore, and as global warming increases the ice-free periods, the hungry bears are more often coming into contact with humans as they search for food in Arctic settlements.

LEFT:
High and mighty
Marking tree trunks conveys a very important fact: the height of the scratch on the tree shows just how tall the bear that made it really is. All but the bravest bears will avoid areas with high marks, in case they come into contact with the scratches' maker.

BELOW:
Bark and bite
When bears meet there is a show of strength. They compare heights, markings and sniff scents, and then there's a display of fangs and growls. This ritualized encounter reduces the chance of violence. Well-matched males, however, will readily fight over a female during the breeding season.

Charge!

If you see this in the wild, you are too close to the bears! Brown bears will charge at a perceived threat, bearing their teeth, bristling their hair and making a lot of noise in an attempt to scare the danger away. It normally works. The charges are mostly mock attacks, with the bear stopping before making physical contact with its target. Few creatures wait around to see what is fake and what is real.

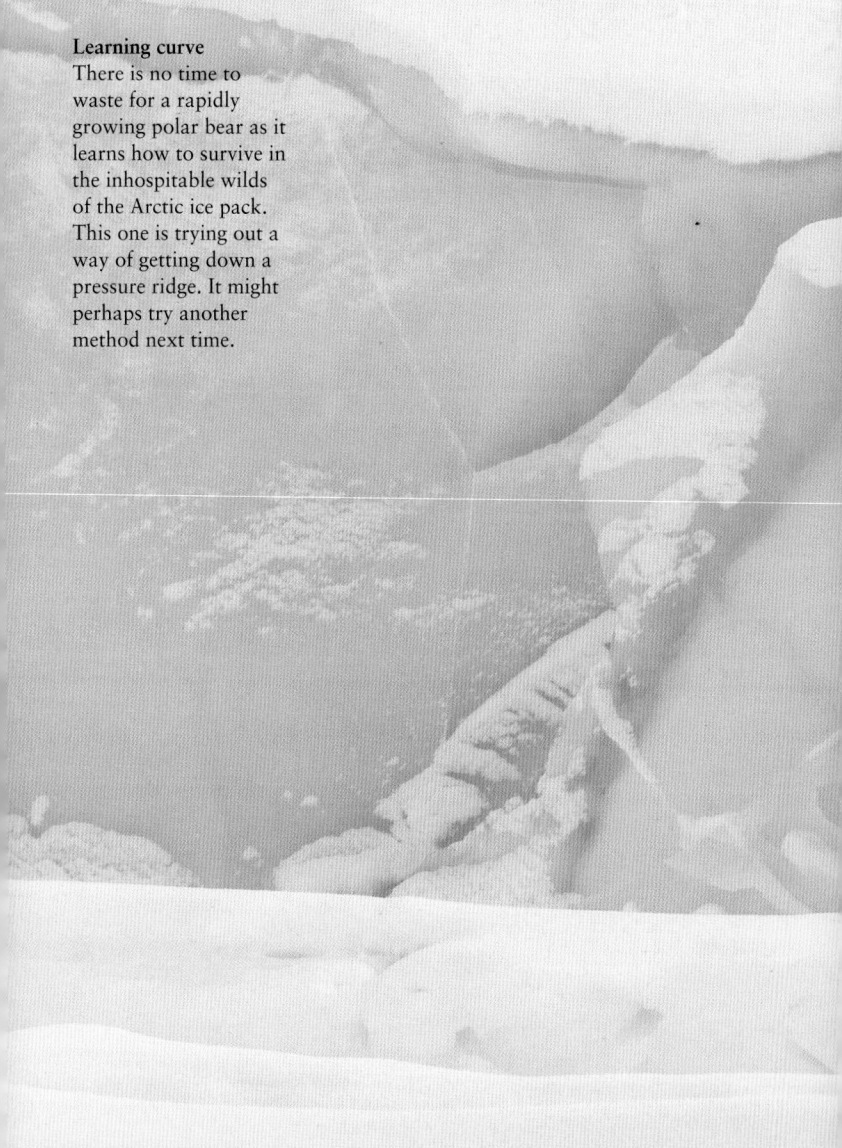

Learning curve
There is no time to waste for a rapidly growing polar bear as it learns how to survive in the inhospitable wilds of the Arctic ice pack. This one is trying out a way of getting down a pressure ridge. It might perhaps try another method next time.

RIGHT:
Nose power
This polar bear is sniffing the ground, perhaps tracking a mate or tracing the scent of prey – or maybe the whiff of a carcass inland that could provide a bit of succour. The bear nose is said to be the most acute of any mammal, judged to be 2000 times more powerful than our sense of smell.

OVERLEAF TOP:
Gripping paws
Polar bears have furry feet that not only help them keep warm but also provide grip on slippery ice. On thin ice, the bears stretch their legs far apart to spread their weight and prevent cracking.

OVERLEAF BOTTOM LEFT:
Swim system
Polar bears use their wide, round forepaws as paddles to haul themselves through the water. The back feet are used as rudders.

OVERLEAF BOTTOM RIGHT:
Long neck
If their white fur were not a big enough giveaway (and that can happen; there are pale members of other species, known as kermodes or spirit bears), then a distinctive feature of the polar bear is the neck, which is consistently longer than in other species. This is an adaptation to reach into holes in the ice to grab prey.

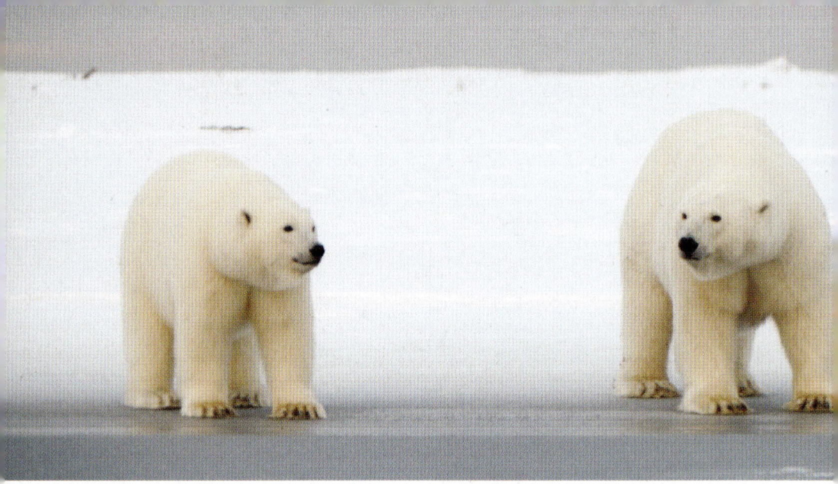

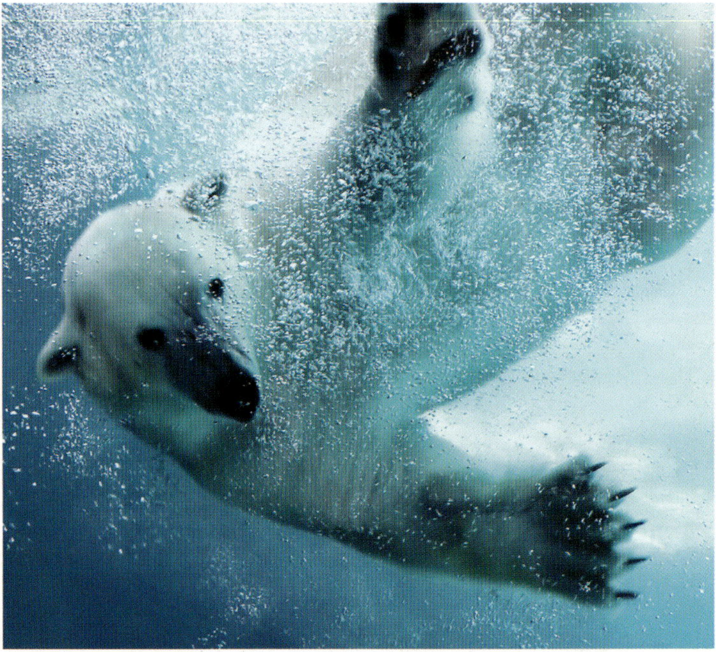

No rest
Bamboo is a very poor source of nutrition, and giant pandas eat more or less constantly, devoting 16 hours a day to ingesting enough calories. It takes just five hours for the meal to pass through the bear's digestive system.

LEFT:
Digging for food
This immensely powerful brown bear is most likely to prey on small animals. It sniffs them out, often as the creatures cower under the leaves or are buried in soil, and digs them up.

ABOVE:
Young game
A brown bear cub climbs a tree trunk as its mother rests. This skill will fade; an adult is unlikely to climb into a tree. Brown bears can climb trees, but their great weight limits them to the lower branches.

Bear encounter
So what do you do if, while taking a stroll, you meet a bear walking the other way? Bears are long-sighted and can see distant objects just as well as a human, and be sure they have smelled and heard you already. At close range, however, the bear becomes rather short-sighted, so you will appear as a bit of a blur. Fast movements will only alarm it, so keep still and identify that you are a harmless creature by speaking in a quiet and constant voice. Now it is time to back away slowly, keeping your eye on the bear at all times – it might still charge. The very best of luck.

Breakfast time
This bedraggled brown bear is looking forward to breakfast. It has lost anything up to half its weight during the winter sleep, and now its survival relies on the chubby salmon that are swarming upstream in spring.

Family

Bears want to be left alone, and to make that crystal clear they spend time marking their territory. First, they leave claw scrapes on tree trunks and will snap a sapling in half to show they have passed by recently. Next, the bears like a good back rub, not just because it no doubt scratches the odd itch but also because it leaves behind thick tufts of hair. Finally, the bear leaves evidence of its scent in well placed dung piles along forest trails and squirts of smelly urine. Additionally, it daubs trees with scent from glands around the body, most notably on the paws and near the anus. Pandas even perform handstands so they can anoint a tree trunk with urine higher up and near nose level. The marks and smells are a bear's calling card, and the local bears will learn the scents of their neighbours and opt to stay out of each other's way. Any strangers arriving in an area are immediately obvious.

Of course, for a short spell in summer the bears are open to receiving visitors for the purposes of breeding. Male bears, known as boars, begin to range widely through the territories of local females, or sows. Both sexes will mate with whichever partners they encounter, but the window of fertility is short so only a single male will be successful in fathering all the cubs. The gestation period for bears is about three months, but the sows use a process called delayed implantation, which postpones the development of foetuses. As a result, cubs are generally born several months after mating.

LEFT:

Born small
A black bear cub takes a walk with its mother. Bear cubs are altricial, which means they are helpless at birth and take a long time to develop motor skills.

ABOVE TOP:

Helping out
Older, larger polar bears have bigger litters than younger bears. They are more experienced and can manage a big family. Older sows have even adopted cubs from litters of younger bears that cannot cope.

ABOVE BOTTOM:

Milk meal
These polar bear cubs were born blind and use their smell to locate the mother's milk supply. She cannot help because she is still fast asleep.

LEFT:

Trio
A litter is often up to four cubs, but only two will survive as the mother struggles to find enough food.

RIGHT TOP:
Summer born
Spectacled bear cubs are born around the same time of the calendar year as those of other species, between December and February. In the Andes, however, that is midsummer, whereas other bears are being born in winter.

RIGHT BOTTOM:
Sniffing the air
Standing tall allows bears to catch the scents hanging on the breeze. This whole family is checking what is out there.

OPPOSITE:
Artificial insemination
The fertility of giant pandas has been hard hit by the reduction in their numbers. In the wild, each adult would have many partners, something that cannot be replicated with captive bears. Chinese breeding programmes have tackled the threat of extinction faced by giant pandas by using artificial insemination and in-vitro fertilization.

PREVIOUS PAGE:
Out and about
After being hidden away in their dens with their slumbering mother, these three-month-old American black bears are enjoying being out and about as the spring weather warms up.

LEFT:
Arctic childhood
As one might expect, life on the Arctic Ocean is not easy. There is much to learn about how to survive and, as a result, these polar bear cubs will stay with their mother for two more winters. Only once she is pregnant again will they move on to start life independent of her and each other.

LEFT:
Creche facilities
In the captive breeding programmes, panda cubs are raised in much larger groups than would happen in the wild. Keepers supply them with food, but are at pains not to interact with the cubs too much because one day the captive bears will be returned to the wild and need to retain a natural fear of humans.

OVERLEAF LEFT:
Leaving their marks
A brown bear cub joins its mother in leaving some fur marks on a trunk. Tufts of long guard hairs will snag in the bark, and their oily coating will carry the bears' scents.

OVERLEAF TOP RIGHT:
Hitching a ride
A brown bear cub takes a rest on its mother's back. This mode of transport is not typical for most bears but looks fun.

OVERLEAF BOTTOM RIGHT:
Watching out
A pair of brown bear cubs steady themselves against their mother's flank as they stand up in a shallow but fast-flowing stream – perhaps to sniff deeply the smell of fish nearby.

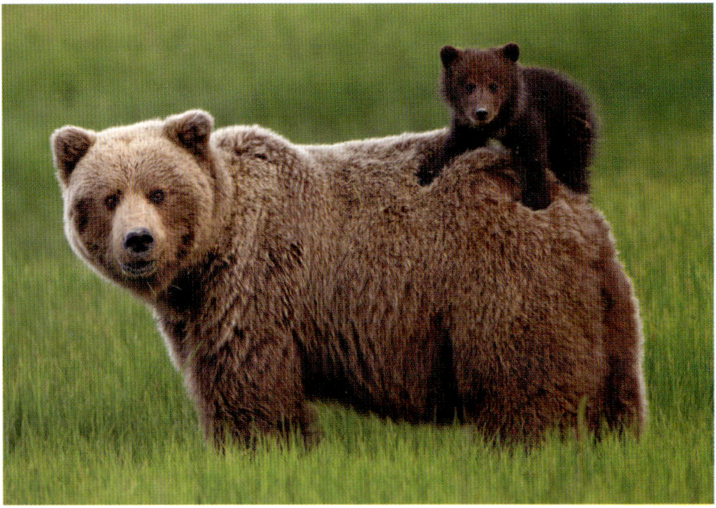

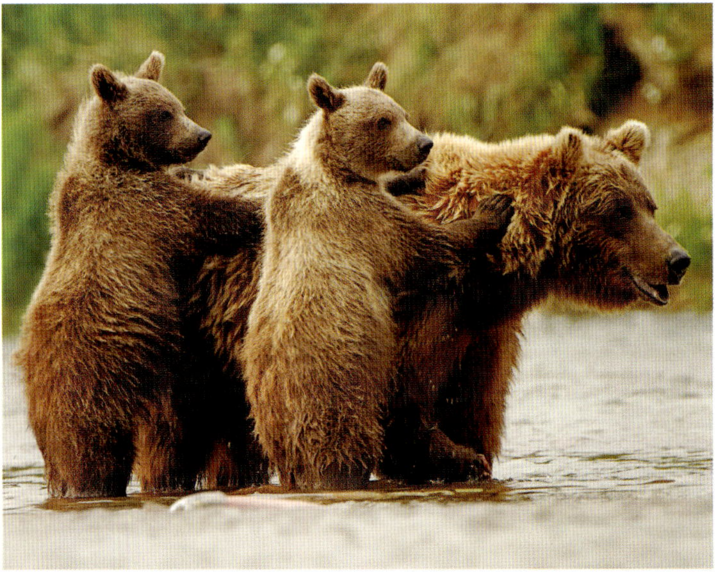

OPPOSITE:

Close cousins
Polar bears are very closely related to brown bears, having diverged from each other less than 100,000 years ago. The brown bear is the older species, with the polar bear developing the appearance of white hair and adaptations to the cold, such as smaller ears and wider feet.

LEFT:

Hidden hybrids
Zoologists suspect that there is a high degree of hybridization in regions close to the Arctic, where polar and brown bears meet. Called grolar bears or pizzlies, the hybrids have a brown coat but also a shorter snout more like the polar species.

Weight loss
A mother bear leading her litter to a feeding site looks somewhat thin. She will have lost up to 40 per cent of her body weight during the long winter sleep.

RIGHT TOP:
Weight difference
A black bear cub weighs around 400g (14oz) when it is born, which is 300 times smaller than the adult. By comparison, a human adult weighs about 18 times more than a baby.

RIGHT BOTTOM:
Taking a position
A sloth bear cub is strong enough to climb on to its mother's back from the age of two. If there are a pair of cubs, each of the twins will always ride in the same position, front or back, each time.

OPPOSITE:
All aboard
Sloth bears are the only species to carry cubs on their backs habitually. Once the cubs are six months old they are too heavy to carry and walk alongside mother.

Fight!
Polar bears mate in summer, when they have had enough time to recover their strength from the last winter sleep, and while there is still time to prepare for the next winter. These polar bear boars are in a show of strength. The winner will live with a sow for a week or two, before heading off to find another mate.

RIGHT TOP:
Exhausting business
Life on the ice is exhausting and this family of polar bears is taking a well-earned summer snooze. When night falls they will go hunting again. The bears' dense coat will keep out the cold very effectively.

RIGHT BOTTOM:
Lazy day
Brown bears are mostly nocturnal, so daytime is an ideal opportunity to take a rest.

OPPOSITE:
Climb time
This tree is an opportunity for these brown bear cubs to have some fun at their mother's expense. She is too big and heavy to haul herself up this trunk.

Watch and learn

Spectacled bears emerge from the den at around the age of three months. They will have to learn fast, because in nine months' time, mother bear might be preparing to give birth to her next cub, and this little bear will be out in the wild looking after itself.

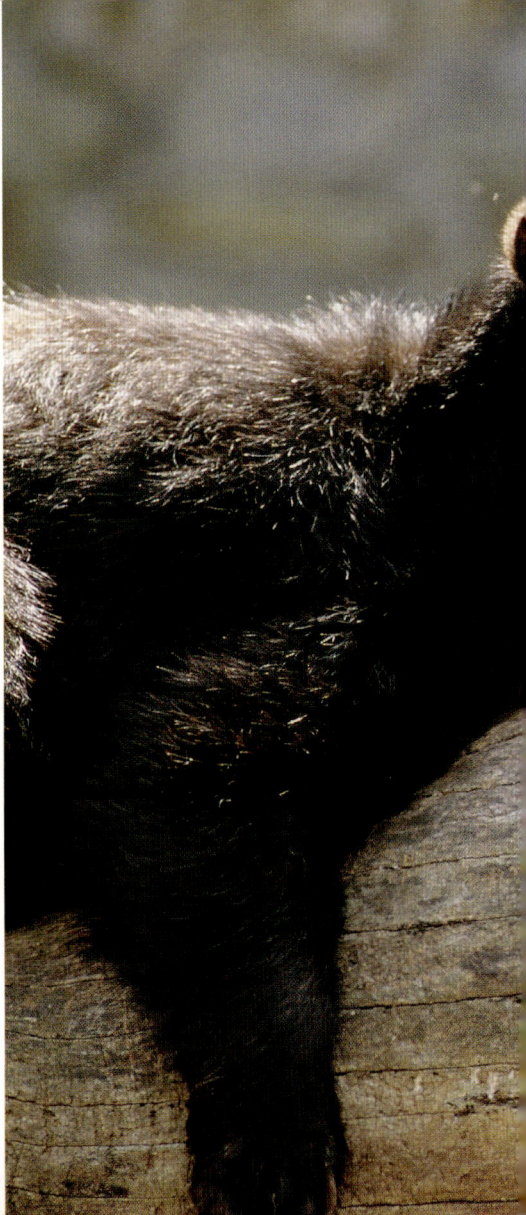

ABOVE:
Offering protection
A polar bear cub is protected by its mother and sheltered beneath her impressive frame. Apart from humans, the only threat to this cub is from another polar bear.

RIGHT:
Play time
A mother and cub living in a zoo have plenty of time to play together. This builds the cub's motor skills. As with humans, it learns through play.

OPPOSITE:

Finding their place
Unlike other bears, which tend to live in more widely spaced areas, giant pandas have a social hierarchy within their more densely populated communities. The older and bigger bears are avoided by the lower-status ones. Entry into the hierarchy starts with the play-fights among panda cubs.

LEFT TOP:

Juveniles
American brown bears stay with their mother for two and a half years. They are driven away when she becomes ready to mate again and starts to attract the attentions of mature males.

LEFT BOTTOM:

Earlier maturity
The Eurasian brown bears, such as this family from Finland, tend to become independent a little earlier than their American cousins. The Eurasian bears are generally a bit smaller and leave their mother at the age of two.

ABOVE:
Maximum load
Bears have four teats, or nipples, and so can feed a maximum of four cubs at once. These brown bears will be weaned off milk from about 18 months. By that time, the mother is not likely to be able to provide milk for all four cubs.

RIGHT:
Follow on
Bears are nomadic creatures, moving daily along well-trodden trails through their home ranges. Once the cubs are strong enough to walk, they will follow their mother and begin the journey of a lifetime.

LEFT:
Protected unit
Cute she might look, but this mother will be capable of extraordinary violence if called upon to defend her family.

OVERLEAF:
Who's the boss?
Playtime for bears is a gentle lesson on how to recognize such signals of dominance as bearing fangs and standing on back legs.

OPPOSITE:

Time to spare
Having eaten a big meal, this polar bear family is taking advantage of some downtime. While not exactly ideal for digestion, this play reinforces the bond between them.

LEFT TOP:

Travel system
A mother brown bear carries her cub on her back across a muddy beach in Alaska. He is, as yet, too small to make his own way easily, especially on this terrain.

LEFT BOTTOM:

Weaning
A pair of hungry panda cubs drink their mother's milk. In their early days, the cubs suckle 14 times a day. They will be weaned at 45 weeks and stay with their mother for 18 months.

LEFT:
Play-fights
The scale is different, and stakes are low, but this play-fight is not far off the real thing. This behaviour represents a form of training for adulthood.

OVERLEAF:
Food search
A family of Alaskan brown bears seems in a hurry to find some food. They will have only recently emerged from the winter den and will be hungry.

RIGHT TOP:
Fishing lesson
A cub looks on as its mother shows it how to catch a fish. Bears will gather at rivers and pools with a large supply of fish.

RIGHT BOTTOM:
Time for action
The Arctic spring is brimming with possibility as this new polar bear family emerge from their winter seclusion.

OPPOSITE:
Working together
This cub has learned fast and is now as alert as its mother. The more self-aware it becomes, the closer it gets to becoming independent.

OVERLEAF:
Self-protection
When it is very young, a bear cub's back legs are much weaker than the forelegs. As a result, they are very unsteady during the first walks, and tend to circle back to where they started. So they never get very far!

RIGHT TOP:
Water bear
A pair of captive juvenile polar bears enjoy a dip in their pool to cool off.

BOTTOM LEFT:
Digest and rest
Polar bears eat much larger meals than other bears, which pick at morsels throughout the day. As a result, polar bears take a bit of rest after eating.

BOTTOM MIDDLE:
Love bite
A polar bear cub tries out its dominance skills on its mother. It is only playing but this is an important way for it to learn.

BOTTOM RIGHT:
Stick together
Polar bear twins are inseparable for two and a half years, but once the time comes to go their separate ways the bears are likely never to see each other again.

Weatherproof
A brown bear mother and cub have everything they need to thrive in the wet and wild forest.

Cubs

Surprisingly, for animals known for their great size, bear cubs start out very small. Newborn cubs are minute compared with their mothers and fathers – much more so than their fellow mammals. The cubs of the giant panda are the record breakers. A panda cub is just 90g (3.17oz) when it is born, which is 1400 times smaller than an adult male.

Bears are born small because they have only a short gestation period, barely a third of the time it takes for a human baby to develop. As a result, they are born almost helpless and are only capable of feeding and sleeping – much like a human baby. Bears evolved this way of staging development as an answer to the big ecological problem they face, one most species confront: how to get enough food in the summer to survive through the next winter. The early altricial, or helpless, phase of a bear's life is completed in winter, while the mother is still dormant. When conditions beyond the den become survivable, the cubs are already fattened up and raring to go as they follow their mother on foraging trips.

The trade-off of this clever technique is that the bears suffer a high mortality rate in the first few years of life. It is relatively easy and low-cost for a mother bear to produce and suckle three or four tiny cubs, but the task of feeding all of them, as they grow rapidly month on month, is harder. Sadly, about a third of all bear cubs die in the first year, and about a quarter of older cubs do not survive long enough to live independently.

OPPOSITE:
No cares
This brown bear cub has everything it needs to grow into one of the world's biggest land predators.

RIGHT:
Long life
This captive polar bear cub has a good chance of living until it is 40 years old. In the wild, it has only a 50 per cent chance of reaching adulthood, but once there could enjoy 30 years of good living.

OVERLEAF LEFT:
Seeing further
A brown bear cub has scaled a stump to get a good look at – and a good smell of – the surroundings.

OVERLEAF RIGHT:
Smelling the way
This cub has been paying attention. It is sniffing a woody stalk for signs that another bear has been past and left its mark.

LEFT:
Calling mother
As it grows, a cub will become bolder and explore further from its mother. If it gets lost, it gives a whining cry, and mother will soon be there.

OVERLEAF:
Claw climber
A young black bear cub uses the claws on all its feet to hang. After a few more weeks, it will be too heavy to hold on. Only sloth bears are capable of this as adults.

LEFT:
Waiting around
After at least three months in the den, up to perhaps as long as six months, a polar bear cub has a full coat and is keen to get itself outside to take its first steps in the strange new world it finds itself in.

BELOW:
Thickening fur
A cub's coat quickly develops the double layer of underfur and guard hairs that are needed to keep it dry, warm and clean, which is especially important as it gets used to the temperatures of its new environment.

OPPOSITE:
Curiosity
A brown bear cub sniffs a tree, learning as much as it can about its world. As an adult, this curiosity will stand it in good stead as it needs to seek out all kinds of food from its dense forest habitat.

TOP LEFT:
Reintroduction
After 18 months in captivity, panda cubs are prepared for reintroduction to the wild. It has been shown that the cubs need to learn social skills from an early age, so they can integrate with wild giant panda populations.

MIDDLE LEFT:
Shelter
Black bear cubs spend a lot of time in trees. This is a good place for them to shelter from the rain.

BOTTOM LEFT:
Nap time
A spectacled bear cub takes a nap during the day. It will be on the move with its mother when night falls.

No human contact
In many cases, the keepers who look after panda cubs in breeding facilities will be dressed in panda costumes, with the same black and white markings. This ensures that pandas are not drawn to humans in the wild and only interact with other giant pandas.

RIGHT TOP:
Thin coat
A very young brown bear cub has yet to develop a thick layer of fur. It will need to dry off and warm up after this dip.

RIGHT BOTTOM:
Easier life
In the warmer, southern parts of the species' range, American black bears find life a little easier. There is no need to pause for long in winter, and so cubs have a better chance of getting the food they need. The biggest problem the bears face is the destruction and fragmentation of their habitat.

OPPOSITE:
Safe space
This black bear is in a place of safety. Trees are a refuge from attack by brown bears, birds of prey, cougars and wolves.

Colour types
As well as black, American black bears can also have cinnamon coats, such as this cub from Minnesota, and less commonly a blue-grey variety found in Alaska and an off-white seen in some Pacific coastal islands.

RIGHT TOP:
Forest life
A brown bear cub is most likely to spend its life in a forest, although its home range might include coastline, grasslands and tundra.

RIGHT MIDDLE:
Unfussy eater
A black bear cub sniffs a lupin. Perhaps it will see how it tastes. To survive, this bear will need to be ready to eat whatever food it can find.

RIGHT BOTTOM:
Bonding
This pair of tiny brown bear cubs will continue growing until they are around 10 years old, even after they become parents themselves.

OPPOSITE:
Sun bear cub
Unlike other bear species, the tropical sun bear has no summer breeding season, and so such cubs as this one are born all year around.

RIGHT:
Copying mother
An eager brown bear cub is copying the adults by grabbing a fish from the river. It might struggle to hold on to it.

OPPOSITE:
Famous bear
This is Knut, a captive-born polar bear who was rejected by his mother at Berlin Zoo in 2007. Knut was raised by keepers and caught the imagination of the public. He went on to have his own video stream and was the subject of a pop song.

OVERLEAF:
The nose has it
Smell is the bear's primary sense and these cubs are learning how to use their powerful noses.

ABOVE:
Visual mark
A panda is busy stripping some bark from a tree trunk to create a visual notation. Often there will be a urine-scent marking at the base of the tree, too.

RIGHT:
Building strength
Climbing is a good way to build muscle, which is important for these young cubs as they quite literally get to grips with their new world.

OPPOSITE:
Stay inside
At seven weeks, this black bear cub still has over a month before it is time to leave its safe and warm den. Leaving too soon would make him vulnerable to all manner of threats, both predatory and environmental.

LEFT:
Swimming lessons
A brown bear will spend a lot of time in water searching for fish, so from an early age cubs are seen to be practising important swimming skills.

ABOVE:
No adults allowed
This brown bear cub will not be disturbed by its mother or any other adult up in these flimsy branches. Older bears are too heavy to reach this part of the tree.

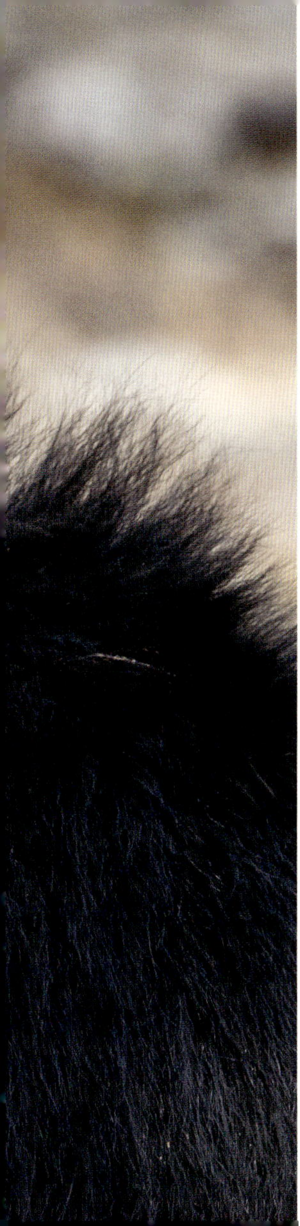

LEFT:
Keeping together
A baby sloth bear gets around on its mother's back. The forest habitat of sloth bears is much more open than for other bear species. The sloth bears evolved this transport system as a way of protecting the cubs from attack by cats and birds of prey.

ABOVE:
Facial features
A tiny spectacled bear is still developing its identity, in that the facial markings used by these bears to tell each other apart have yet to fully form.

RIGHT TOP:
Precarious existence
Most species of bear are not in imminent danger of extinction, but in many areas of the globe local populations have a precarious future thanks to poaching and habitat destruction.

RIGHT BOTTOM:
Animal ambassador
The most endangered bear is the giant panda, which is also the most easily recognized. The distinctive features of the panda made it the perfect choice to be the symbol for the world's conservation efforts to protect threatened animals. And it was one of the first successes. Thanks to breeding programmes, the giant panda has been taken off the critical list.

OPPOSITE:
Taste the difference
If in doubt, give it a lick. This bear is investigating everything it can about the forest habitat and the bears that live in it.

The future
What does the future hold for this black bear cub and its cousins across the world? A major threat will be climate changes that shift or reduce the bears' habitats. Will these mighty creatures be able to adapt to the changes?

Picture Credits

Alamy: 8 (Andrew Digby), 11 top (agefotostock), 11 bottom (David Kilpatrick), 12 (Mark Waugh), 13 (charles nolder), 16 top (Bill Lea/Dembinsky Photo Associates), 20 (Vince Burton), 32/33 & 36 (Arco Images), 48 bottom (Ken Gillepsie Photography), 52 (Anan Kaewkhammul), 53 (Nature Picture Library), 54 (A+J Visage), 55 (Nature Picture Library), 62/63 (Lou Linwei), 72/73 (Paulette Sinclair), 74 top (Ilene MacDonald), 78/79 (Galaxiid), 90 top (infocusphotos.com), 90 middle (Minden Pictures), 91 (Arco Images), 99 (All Canada Photos), 102/103 (Arco Images), 105 (Keren Su/China Span), 106/107 (Minden Pictures), 108/109 (Nature Picture Library), 117 (imageBROKER), 122/123 (Accent Alaska.com), 124/125 (Russell Millner), 126/127 & 128/129 top (Nature Picture Library), 134/135 (All Canada Photos), 136/137 (Renato Granieri), 144/145 (Arco Images), 148/149 (Helen Nicholson), 154/155 (Design Pics Inc), 162/163 (Juniors Bildarchiv), 164 (Richard Mittleman/Gon2Foto), 166 (Gang Liu), 169 (Russell Millner), 201 top (WILDLIFE GmbH), 201 bottom (blickwinkel), 214 bottom (Don Johnston_WC), 219 (Ronald Wittek)

Dreamstime: 14 (Rinus Baak), 15 (Holly Kuchera), 21 (Volodymyr Byrdyak), 26 (Lori 0469), 28/29 (Graham Prentice), 40/41 (Micha Klootwijk), 48 top (Andrey Gudkov), 49 (Anatoly Frumgartz), 50/51 (Theeravat Boonnuang), 64/65 (Vladimir Cech), 68 (Glenn Nagel), 76/77 (David Kennedy), 82 (Amanda Mortimer), 83 top (Mikael Males), 84/85 (Gabriela Adamkova), 133 (Sergey Uryadnikov), 141 top (Belovodchenko), 158/159 (Sjors Stans), 160 top (Outdoorsman), 160 bottom (Genta27), 168 (Ricochet69), 172/173 (Natalia Golovina), 178/179 (Tony Campbell), 180 top (Petr Simon), 181 (Mikael Males), 184/185 top (Taraileso09), 184 bottom (Sergey Sklezniev), 188 (Sergey Uryadnikov), 196/197 (Rstrick2), 199 (Amanda Mortimer), 202/203 (Hupeng), 204 bottom (Jianqing Gu), 206/207 (Jocrebbin), 212/213 (Lucaar), 214 top (Ondrej Prosicky), 217 & 221 (Sergey Uryadnikov)

FLPA: 24/25 (Matthias Breiter), 27 top (Jack Chapman), 98 (Richard Garvey-Williams/Nature in Stock), 104 bottom (Jurgen + Christine Sohns), 118/119 (Matthias Breiter), 142 top (Gerard Lacz), 167 bottom (Paul Hobson), 175 top (Ingo Arndt), 180 bottom (Andre Gilden/Nature in stock), 194/195 (Jurgen + Christine Sohns), 200 (IMAGEBROKER, MARKO KANIG)

FLPA/Biosphoto: 43 bottom (Paul Souters), 44/45 (Jean-Jacques Pangrazi), 80/81 (Sylvain Cordier), 86/87 (Meril Darecs + Manon Moulis), 104 top & 151 (Fabrice Simon), 157 (Sylvain Cordier), 185 bottom right (Patrick Kientz), 201 middle, 216 & 222/223 (Sylvain Cordier)

FLPA/Minden Pictures: 7 (Suzi Eszterhas), 18/19 (Donald M. Jones), 27 bottom (Sergey Gorshkov), 37 top (Kevin Schaefer), 37 bottom (Cyril Ruoso), 56/57 (Thomas Marent), 58 top (Ingo Arndt), 83 bottom & 88/89 (Sean Crane), 96 (Sergey Gorshkov), 150 (Marion Volborn, BIA), 151 top (Ingo Arndt), 156 top (Suzi Eszterhas), 167 top (Tim Fitzharris), 175 bottom, 215 (Suzi Eszterhas)

Getty Images: 16 bottom (David Palmer), 34/35 (Tambako the Jaguar), 38/39 (Paul Souders), 42 (Hans Strand), 46/47 (KeithSzafranski), 59 (Mike Powles), 94/95 (Paul Souders), 100/101 top (Mint Images - David Schultz), 100/101 bottom (Chase Dekker Wild-Life), 111 (Ralph Lee Hopkins), 112/113 (Juan Carlos Vindas), 116 (Barrett Hedges), 138 (LOIC VENANCE), 140 (ANDREY SMIRNOV), 156 bottom (Joe McDonald), 165 (Janine Schmitz), 185 bottom left (49pauly), 211 (JOHN MACDOUGALL), 218 (Ignacio Palacios)

iStock: 142 bottom (N8turGrl)

Shutterstock: 6 (FloridaStock), 10 (Molly NZ), 17 (Jim Cumming), 22/23 (Ondrej Prosicky), 30/31 (Dennis W Donohue), 43 top (FloridaStock), 58 bottom (clkraus), 60/61 (Hung Chung Chih), 66 top (Nagel Photography), 66 bottom (Photocech CZ), 67 (Brian Upton), 69 (krechet), 70 (ArCaLu), 74 bottom (Bryan Faust), 75 (Martin Rudlof Photography), 90 bottom (Andrea Izzotti), 92/93 (Thomas Hulik ART point), 110 (Sergey Sklezniev), 114/115 (Martin Rudlof Photography), 120 & 121(Erik Mandre), 128 bottom (Sylvie Bouchard), 129 bottom (FloridaStock), 130/131 (Fernon Archilla), 132 (Canon Boy), 141 bottom (Belovodchenko Anton), 143 (kiszon pascal), 146/147 (Sergey Uryadnikov), 152 (sibsky2016), 153 (Nicholas Hunter), 161 (Erik Mandre), 170/171 (Diego Cottino), 174 (Lamberrto), 176/177 (Giedriius), 182/183 (jadimages), 186/187 (Martin Rudlof Photography), 190/191 (Ger Basma Photos), 192 (godi photo), 193 (Sergey Uryadnikov), 198 (Shvaygert Ekaterina), 204 top (Volodymyr Burdiak), 205 (Holly Kuchera), 208 top (GobyOneKenobi), 208 middle (Geoffrey Kuchera), 208 bottom (Volodymyr Burdiak), 209 (Ryan Ladbrook), 210 (Egor Vlasov), 220 top (Agnieszka Bacal), 220 bottom (shejian)